SUPER-DANGEROUS ANIMALS

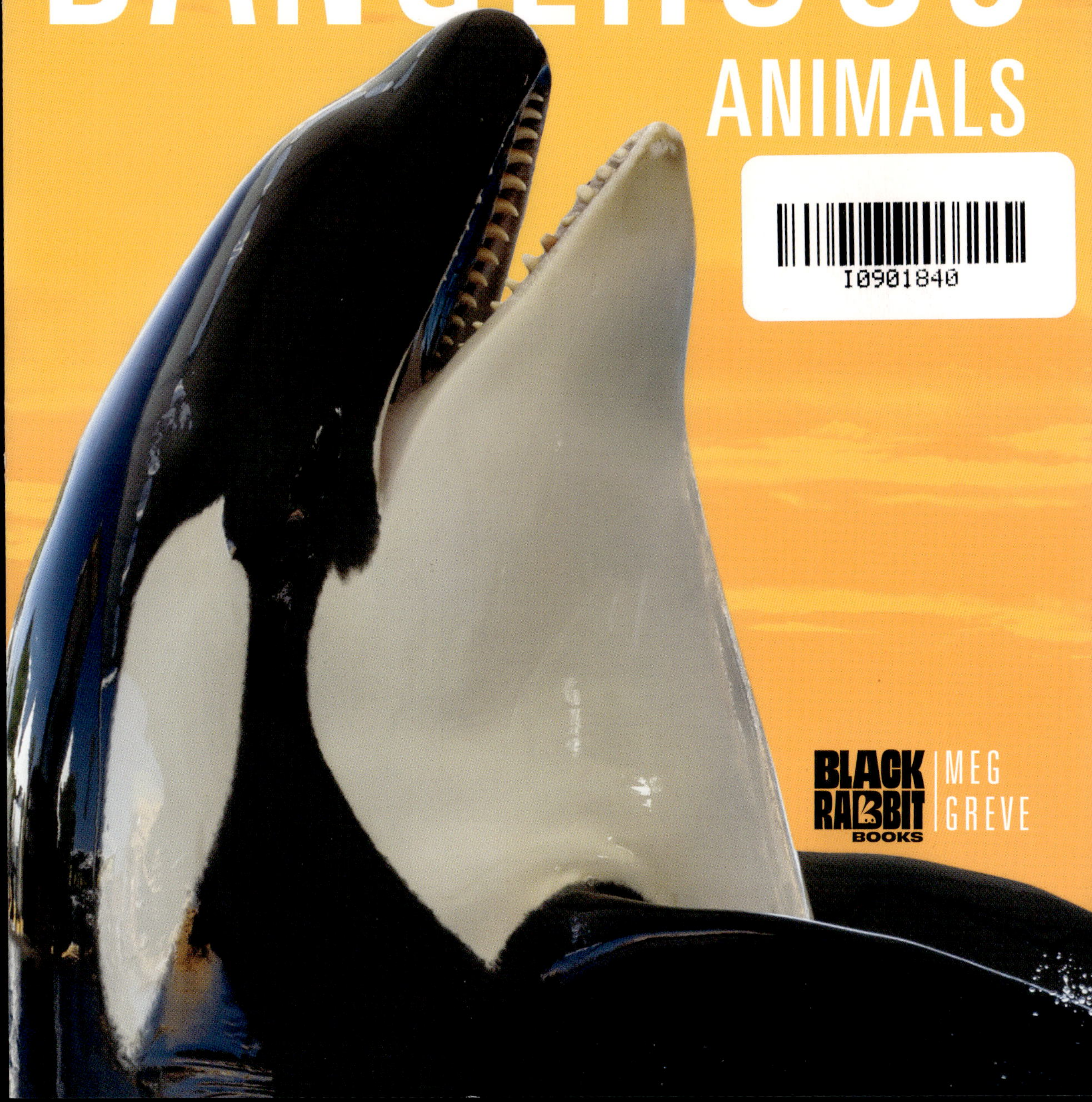

BLACK RABBIT BOOKS

MEG GREVE

TABLE OF CONTENTS

1

The Cassowary

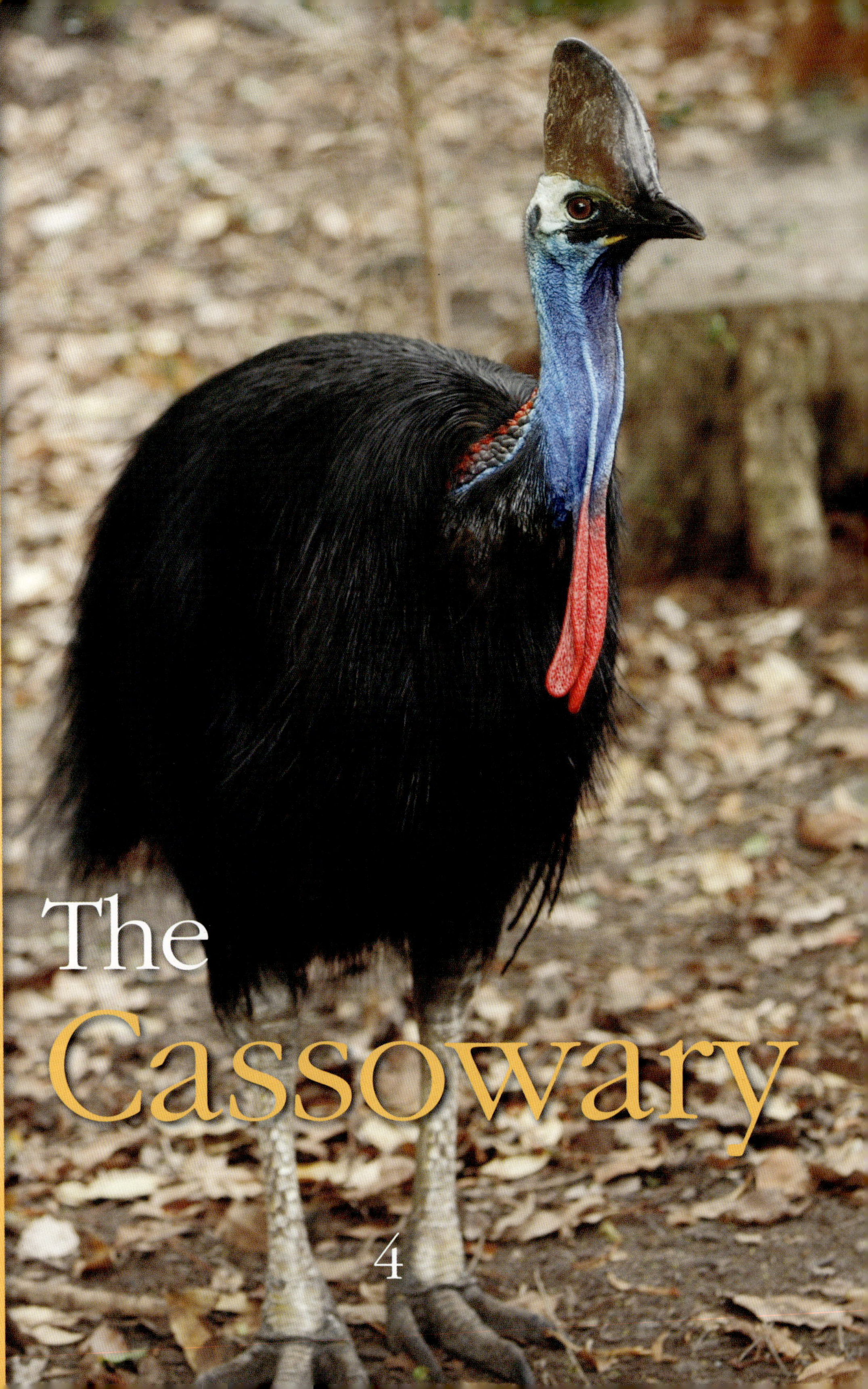

A cassowary is a large bird. It has beautiful coloring. But it is also dangerous. It has a powerful kick. The blow has even killed humans.

This bird can run up to 30 miles (48 kilometers) per hour. It can also jump up to 5 feet (1.5 meters) high. But it cannot fly.

The cassowary eats fruit, small animals, and bugs. It has a casque at the top of its head. This is like a helmet. It helps the bird push through thick rainforest plants.

Think About It

Cassowaries have bright colors on their necks. How do you think this helps them survive?

2

The Hippo

When people yawn, it is a sign of being tired. When a hippo yawns, it is a warning to stay away!

Hippos have giant mouths and strong jaws. Their long tusks help them fight. Their skin is 2 inches (5 centimeters) thick. It is too deep for sharp claws.

Hippos move fast in the water. They can be aggressive. They fight over food and space. They attack lions, leopards, and crocodiles. And they often win!

Did You Know?

Hippos are big. They are the second largest land animal.

3

The Lion

A lion's roar can be heard 5 miles (8 km) away. But this wild cat is quiet while hunting. Its prey might not know it is in danger until it is too late.

Lions are big cats. They live in a pride. Often, the female lions do the hunting. They are smaller and faster. And they hunt as a group.

The cats sneak up on prey. They surround the animal so it can't escape. Then they attack in a burst of speed. Lions can catch antelope, crocodiles, pythons, and hippos.

Did You Know?

Lions can live up to 14 years.

4

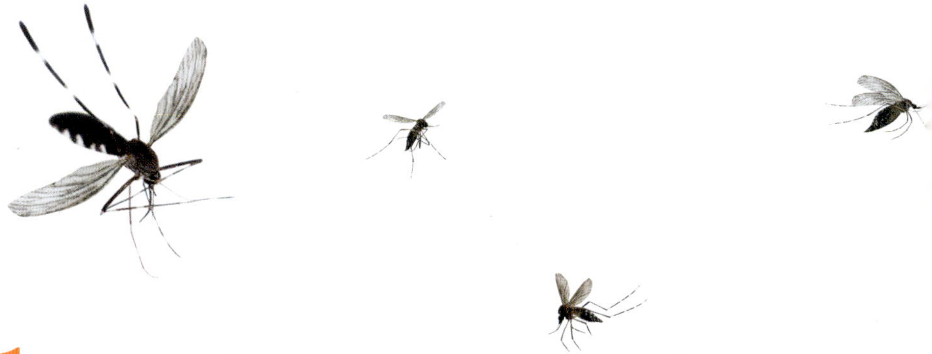

The Mosquito

Mosquitoes are tiny but mighty insects. They can be dangerous. They drink blood from animals and humans. They also carry diseases. They kill more than 3 million people a year.

These bugs sense odor and body heat. They have a mouth part called a **proboscis**. It sticks into the skin and sucks blood.

Mosquitoes only live up to three weeks. It is not a long life. But females lay hundreds of eggs in that short time.

Think About It

Why is it important for you to protect yourself from mosquito bites?

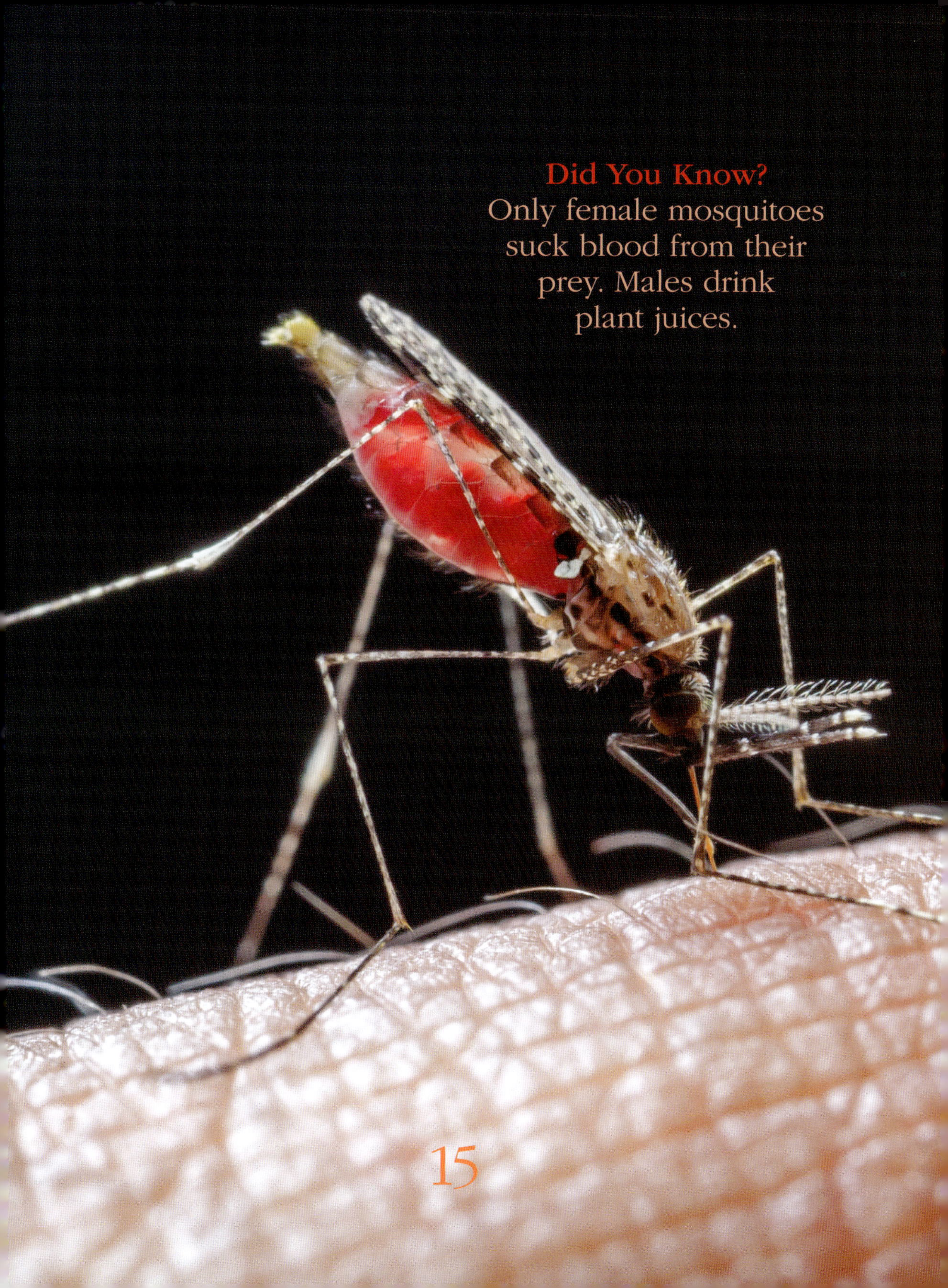

Did You Know?
Only female mosquitoes suck blood from their prey. Males drink plant juices.

5

The Orca

Orcas are called killer whales for a reason. They live in a group called a pod. Each pod of about 30 whales is led by a female.

When the pod is hungry, they hunt together. Orcas are smart. They make clicking sounds to communicate. They plan their attacks. Sometimes, they herd fish into a group. Then they strike the fish with their tails. Other times, they create waves to knock prey off floating ice.

Favorite meals include sea lions, walruses, sea elephants, turtles, penguins, and squid. Orcas even attack bigger whales.

Think About It

Orcas hunt together. How does this help the members of their pod?

6

The Wolverine

What looks like a bear and smells like a skunk? A wolverine! These creatures are related to weasels but stronger and larger.

Wolverines are bold predators. They have an excellent sense of smell. They are powerful with strong teeth and claws. They are built to tear into frozen meat.

These animals are **scavengers**. They usually eat dead animals killed by other predators. But they can also kill prey much larger than themselves, like moose and caribou. They will also eat berries and plants.

Did You Know?
Wolverines may travel 15 to 18 miles (24–29 km) each day in search of food.

MORE TO EXPLORE

WHERE IN THE WORLD?

WOLVERINES
North America

HIPPOS
Middle parts of Africa

LIONS
Southern parts of Africa

MOSQUITOES
All over the world except Antarctica

ORCAS
All oceans

CASSOWARIES
Papua New Guinea, Indonesia islands, and Australia

MORE TO EXPLORE

FANTASTIC FACTS

A mosquito's wings beat up to 1,000 times a second.

Cassowaries are strong swimmers.

Orcas have about 45 sharp teeth used for ripping and tearing their prey.

Lions are the only cats that live in groups.

Hippos can sleep in the water without drowning.

Wolverines can climb trees and are also able to swim.

MORE TO EXPLORE

COOL COMPARISONS

Mosquitoes
0.4 inches
(1 cm) long

Wolverines
1.5 feet (0.5 m) tall
3.5 feet (1.1 m) long

Lions
4 feet (1.2 m) tall
10 feet (3 m) long

Hippos
Up to 5 feet (1.5 m) tall
16.5 feet (5 m) long

Cassowaries
5.5 feet (1.7 m) tall

Orcas
26 feet (8 m) long

MORE TO EXPLORE

RESOURCES

Glossary

aggressive (uh-GRES-iv) Showing a readiness to fight, argue, or attack.

casque (KASK) Protection for the head.

pride (PRIED) A group of lions living together.

proboscis (pre-BOS-es) A long, thin tube that is part of a mosquito's mouth.

scavenger (SKAV-in-jer) An animal that feeds on dead plants and animals.

tusk (TUSK) A long, large tooth.

Read More

Clarke, Ginjer L. *Tiny Terrors!: The World's Scariest Smallest Creatures.* New York: Penguin Young Readers, 2022.

Deadliest Animals on the Planet. Washington, DC: National Geographic Kids, 2023.

Mather, Charis. *Killer Critters.* Minneapolis: Bearport Publishing Company, 2024.

Index

TOP RANK is published by Black Rabbit Books, P.O. Box 227, Mankato, MN, 56002.

• Top Rank is an imprint of Black Rabbit Books. • Edited by Alissa Thielges • Designed by Danny Nanos • Photographs © Dreamstime/Julesuyttenbroeck, 8, Thor Jorgen Udvang, cover; Getty Images/CraigRJD, 4, doug4537, 14, 21, WLDavies, 10; iStock/Tammy Wolfe, 16; Shutterstock/2630ben, 11, 7th Son Studio, 15, Balora, 23, bigkai, 23, erdistudio, 23, Eric Isselee, 12, 21, Foto 4440, 18, Gallinago_media, 20, harkani, 23, Jamen Percy, 20, Johan Larson, 5, 21, Kwangmoozaa, 13, Mari Swanepoel, 9, Michal Ninger, 19, 21, Mike Price, 2, Nynke van Holten, 20, PhotocechCZ, 21, Tereza Tothova, cover, Tomas Drahos, 6–7, Tory Kallman, 17, 21, Viktoria Suslova, 23, white whale, 23 • Printed in China

Library of Congress Cataloging-in-Publication Data: Names: Greve, Meg, author. | Title: Super-dangerous animals / by Meg Greve. | Description: Mankato, MN: Top Rank is an imprint of Black Rabbit Books, [2025] | Series: Super-incredible animals | Includes bibliographical references and index. | Audience: Ages 8–11 | Audience: Grades 2–3 | Identifiers: LCCN 2024017571 | ISBN 9781645823773 (library binding) | ISBN 9781645823995 (paperback) | ISBN 9781645824213 (ebook) | Subjects: LCSH: Dangerous animals—Juvenile literature. | Classification: LCC QL100 .G74 2025 (print) | DDC 591.6/5—dc23/eng/20240517 | LC record available at https://lccn.loc.gov/2024017571